NATIONAL AUDUBON SOCIETY
POCKET GUIDES

National Audubon Society

The mission of the NATIONAL AUDUBON SOCIETY *is to conserve and restore natural ecosystems, focusing on birds and other wildlife for the benefit of humanity and the Earth's biological diversity.*

We have nearly 600,000 members and an extensive chapter network, plus a staff of scientists, lobbyists, lawyers, policy analysts, and educators. Through our sanctuaries we manage 150,000 acres of critical habitat.

Our award-winning *Audubon* magazine carries outstanding articles and color photography on wildlife, nature, and the environment. We also publish *American Birds,* an ornithological journal, *Audubon Activist,* a newsjournal, and *Audubon Adventures,* a newsletter reaching 500,000 elementary school students. Our *World of Audubon* television shows air on TBS and public television.

For information about how you can become a member, please write or call the Membership Department at:

NATIONAL AUDUBON SOCIETY
700 Broadway, New York, New York 10003
(212) 979-3000

NORTH AMERICAN WATERFOWL

Text by Richard K. Walton

Alfred A. Knopf, New York

This is a Borzoi Book
Published by Alfred A. Knopf, Inc.

Published in the United States by Alfred A. Knopf, Inc.,
New York, and simultaneously in Canada by Random House
of Canada Limited, Toronto. Distributed by
Random House, Inc., New York.

Prepared and produced by Chanticleer Press, Inc.,
New York.
Printed and bound by Dai Nippon Printing Co., Ltd., Tokyo.

Published February 1994
First Printing

Library of Congress Catalog Number: 93-21255
ISBN: 0-679-74924-1

Contents

How to Use This Guide

From elegant white swans to brightly colored ducks to familiar Canada Geese, waterfowl are a varied and fascinating group of birds. This guide is designed to help you identify these birds in their natural habitats and learn about their habits, life cycles, and history.

Coverage   The guide features 74 species of North American birds associated with fresh or salt water. In addition to the swans, geese, and ducks that are traditionally termed waterfowl, it includes a variety of loons, grebes, cormorants, and rails that may be encountered in the same habitats. The birds are presented in the American Ornithologists' Union's taxonomic order.

Organization   There are three parts to this guide: introductory essays; color plates and accompanying descriptions; and appendices.

Introduction   The essay called "Birdwatching" discusses some of the reasons that the pastime is so popular and offers tips on equipment that can enhance the experience. "Identifying Birds" outlines the characteristics you should notice when you look at a bird. Finally, "Watching Waterfowl" tells you where and when to best observe these birds and describes the families represented in this guide.

6

The Birds The first part of this color section presents the birds in their natural habitats. Facing each photograph is a description of the species, including a short discussion of its unique or unusual traits; the bird's adult size and field marks; and its voice, habitat, and range. To supplement the range statement, there is a map on which breeding and winter ranges are indicated by diagonal hatching. Where there is an overlap, or wherever a species occurs year-round, the ranges are superimposed. In rare cases, a species occurs so infrequently or in such small numbers that no range map is provided. The small silhouette accompanying each species account is designed as an aid in identification. These silhouettes represent general body types but do not necessarily indicate the subtle variations that can occur among species of the same family.

Breeding range

Winter range

Permanent range

The second part of the color plates presents some waterfowl in flight, with descriptions of characteristics that help to identify the birds on the wing.

Appendices Following the photographs are a drawing of a representative bird and another of a wing, with field marks labelled. The glossary defines additional terms that may be unfamiliar.

7

Birdwatching

Birdwatching is a fast-growing and enormously popular pastime. Estimates suggest that between 60 and 80 million people are involved with "birding" to some extent. Many of these enthusiasts are backyard birdwatchers, who enjoy feeding their feathered neighbors in return for the opportunity to observe the colorful diversity of local bird life throughout the year. At the other end of the birdwatching spectrum are those whose birding takes them far afield. These birders travel regularly to regional parks and conservation areas and, when time and finances allow, to birding "hot spots" throughout North America. Many birders join in organized bird walks, censusing projects, and fund-raising activities for their local conservation organizations. Typical of this group are the more than 42,000 birders who participate in over 2,000 National Audubon Society–Leica Christmas Bird Counts and National Audubon Society Birdathons, held annually throughout North America. Even more people enjoy birds as a complement to other activities, such as gardening, hiking, jogging, boating, and fishing.

Why do so many people watch birds? There may be as many answers to this question as there are birders. In general, however, birds are diverse, colorful, readily observable, and

8

reasonably easy to identify. Another important attraction may be the relative simplicity of birding. Once you are equipped with a bird guide and binoculars, you are ready to begin. Yet this simple beginning may contain the seeds of a lifelong passion. Whatever direction your new-found interest takes, it won't be long before you can enjoy at least a familiarity with the common species in your area.

Binoculars A good pair of binoculars will make birdwatching enjoyable, while binoculars that have poor optics or are damaged will make birdwatching frustrating and even painful. If you already own binoculars, you should "field test" them to see if they are suitable for observing birds. Check the binoculars for clarity of image and ease of focus. Problems may include multiple images, difficult focusing, and image distortion. Essentially, your binoculars should be easy to focus and offer comfortable viewing; eyestrain may indicate that the optics are poor or damaged.

While some beginning birders may be fortunate enough to own a good pair of binoculars, others will want to invest the time and money it takes to learn about and purchase a good pair. There are books on how to buy binoculars, and you may wish to consult one, but a few basic facts will get you started. Specifications for binoculars are stated in a form that includes two numerals separated by a multiplication sign: for example, 7×35. The first numeral refers to overall magnification; the second, to the diameter (in millimeters) of the objective lens (the lens closest to the object you are observing). So, in this example, the binocular magnifies the image seven times and has an objective lens that is 35 millimeters wide. If you divide the second numeral by the first, the result is what is called the "exit pupil." The exit pupil in this example is 5. This number gives you a relative idea of how much light reaches the eye. A 10×40 binocular would have an exit pupil of 4, indicating that less light will reach the eye.

Beyond these specifications, comparisons tend to become more difficult. An inexpensive pair of binoculars might have a high exit pupil but poor optics. On the other hand, some excellent binoculars have a relatively low exit pupil but provide a clear, bright image because of their exceptional optics and lens coating. Fine binoculars combine high-quality optics with durable construction.

The old saw "You get what you pay for" definitely applies to the binocular market. At publication, top-grade binoculars cost anywhere from $600 to $1,000. Very acceptable mid-range binoculars, such as several Bausch and Lomb models endorsed by the National Audubon Society, may also be bought for $200 to $500. You may find a good pair of binoculars for less than that, but they are much more likely to have average optics as well as less overall durability. It is wise to shop around, however. For example, one top-rated binocular lists at $1,500 but is widely available for $900.

Spotting Scopes — Certain groups of birds, such as raptors and shorebirds, are better observed with spotting scopes, which are similar to telescopes mounted on tripods. These scopes provide higher magnification than most binoculars; the tripods provide the stability necessary with increased magnification. As is true of binoculars, a wide assortment of scopes is available. Depending on the quality of their optics and housings, prices range between $200 and $1,000. The tripod, an important added consideration, should be chosen for its stability and portability. Avoid spending so much on a scope that you have to settle for a "make-do" tripod. An acceptable scope and tripod will cost about $300.

Identifying Birds

More than a few beginning birdwatchers have had the experience of spending time with a seasoned birder who correctly identifies every flitting feather with miraculous ease. Despite outward appearances, the beginner has not witnessed a miracle, and the expert has no superhuman abilities. His or her expertise has developed over time. Given enough time and practice, you can develop similar skills.

Range
An important first step in identifying birds correctly is knowing what birds to expect, as well as when and where to expect them. A state bird list, available from your local chapter of the National Audubon Society, will be useful. This Pocket Guide will also help you narrow down the possibilities. Browse through the guide, noting the ranges of the various species, including their seasonal distributions. This will help you to decide if and when a particular species is likely to be in your area. (You may wish to mark pages describing birds that often occur in your vicinity.) The Pocket Guide series, because it is organized by types of birds—*Birds of Prey* and *Waterfowl,* for example—allows you to further reduce the possibilities. Your first goal is to know if and when you might expect to see any given bird in your area.

Size A notion of the overall size of a bird is a critical component of identification. Each species account in this guide includes the body length (from tip of bill to tip of tail) of the bird. Perhaps just as useful are comparative descriptions such as "sparrow-sized" or "crow-sized." Field experience with benchmark species such as the Song Sparrow, American Robin, Blue Jay, American Crow, and Canada Goose will prepare you to use the general sizes of these familiar species in a comparative way to identify unfamiliar birds.

Shape Some birds can be identified by their shapes alone. In the field, under various light conditions, a silhouette may be all that is discernible. As you gain experience, you will be able to use overall proportions as well as the specific shapes of bills, heads, and tails to confirm a bird's identity.

Color and Pattern The colorful feathers of many birds are important identifying characteristics. More often than not, however, it is a combination of color and pattern that confirms a bird's identity. In North America, a robin-sized songbird with an all red head and body, and black wings and tail is a Scarlet Tanager. The Scarlet Tanager is a straightforward example, but the color and pattern of feathers on other species may be quite complicated. In order to deal with this fact, experts use a somewhat specialized vocabulary. Terms such as crown, eye-stripe, eye-ring, and undertail are readily understandable. Other terms, such as mandible, speculum, undertail coverts, and flight feathers, require explanation. To become familiar with what such terms mean, refer to the illustrations and glossary at the back of this guide.

Behavior Behavioral clues are also useful for identification. Some birds habitually flick their tails, others scratch among the leaves, and still others bob their heads up and down. Some ducks dive and others dabble. There are hawks that hover and hawks that stoop. Some large water birds fly with head and neck outstretched, while others draw in the neck and head. Typical behavior is mentioned in the general description or under "Identification" in each species account.

Voice Beginners would do well to spend at least some time learning the songs and calls of birds. Virtually all North American birds can be identified by voice alone. Each species account in this Pocket Guide includes a section on "Voice." Both descriptive terms and phonetic clues are given. Start your learning with familiar birds that you hear regularly, and don't be shy about modifying or adding to the ideas given here with your own characterizations of birds'

14

songs. These calls and songs are not only enjoyable to learn, but they can also be very useful in places such as dense woodlands and extensive thickets. A variety of commercial products are available to help you learn bird songs.

Gestalt Experienced observers use any or all of the clues mentioned above, often in combination with a process of elimination, to identify birds. At times a flick of the tail, a single call note, a flash of color, or a familiar silhouette is enough to confirm an identification. More often a combination of factors reveals the species. At times, birdwatchers refer to the "jizz," or gestalt, of a bird—some overall look or impression that may be difficult to analyze but somehow suggests the given species. Most likely, jizz or gestalt is simply a combination of the various factors outlined here. Ultimately, it is time in the field that will enable you to develop your identification skills.

Watching Waterfowl   The waterfowl included in this guide are a diverse and
popular group of birds. While observers' interests will no
doubt vary with their perspectives—be they birdwatchers,
photographers, hunters, or the public at large—waterfowl
hold a special attraction for us all. Some species are familiar
sights in local parks and ponds, while others are well known
for making extraordinary migrations or over-wintering in
spectacularly large flocks. Many waterfowl are appreciated
simply for their colorful plumages. Learning more about
this group of birds is a project that promises many
enjoyable hours in the field.

The Birds   Traditionally, the term waterfowl refers to the various
species of ducks, geese, and swans. This volume presents
over 50 species of waterfowl that occur in North America. In
addition, it includes a variety of North American loons,
grebes, cormorants, and other birds that, more often than
not, you will encounter in the same habitats as waterfowl.
Thus this guide is designed to be a useful reference for
identifying a somewhat more inclusive group of birds.

A broad coverage of wetland birds in general can be found
in two companion guides. Freshwater wetland species are
treated in the *National Audubon Society Pocket Guide to*

Familiar Birds of Lakes and Rivers, and birds associated with saltwater appear in the *National Audubon Society Pocket Guide to Familiar Birds of Sea and Shore.*

Where and When to Look   The common denominator for all the species shown in this guide is water, whether it is freshwater ponds and marshes or saltwater marshes and estuaries. Although locating some waterfowl, such as Mallards and Canada Geese, requires little searching, finding other species will involve some planning. During the migratory periods of spring and fall, waterfowl may show up almost anywhere. Local ponds, lakes, and reservoirs are good places to start looking. In spring, you may want to spend some time searching along the edges of flooded rivers or in wet fields. It is surprising how small a puddle some duck species require.

A wide variety of ducks, including Gadwall, Blue-winged and Green-winged teal, American Wigeon, and Northern Pintail, show up inland across much of North America. Other species migrate along the coast, so a trip to the seashore normally provides an opportunity to view a variety of waterfowl. Ocean storms, coinciding with the migrations, make for spectacular viewing at times. Loons, grebes, and ducks are among the birds driven by such storms.

17

During the breeding season, lakes, marshes, prairie potholes, and the Arctic tundra play host to most of our nesting waterfowl. While a trip to the Arctic is not possible for many of us, a visit to a local pond or wildlife refuge is likely to provide familiarity with one or two species as they go about the business of raising their young. During the winter months, many of these birds take advantage of the havens provided by wildlife refuges, especially those along both coasts and in southern sections of the United States. Locating one or two such refuges in your area should enhance your viewing opportunities.

Getting Started
Most waterfowl are reasonably easy to observe, although they may become more wary when the hunting season is in progress. A spotting scope is useful for identifying waterfowl, particularly those species grazing in fields or dabbling and diving on large bodies of water. Consult the section called "Birdwatching" for additional tips.

Conservation
Waterfowl are among the most vulnerable of all species to human-caused threats. The continuing destruction of wetlands and pollution of estuaries and seashores throughout the hemisphere jeopardize the nesting and wintering grounds, as well as migratory stopover points of

18

countless species. Taking an interest in watching these unique creatures may inspire you to participate in the efforts of the National Audubon Society, and others, to protect bird habitats.

The Families   Birds have been arranged by biologists in groups called families. Knowing the characteristics of the families represented in this guide will help you in identifying the birds you see.

Loons   Loons (family Gaviidae) are large, highly aquatic birds that dive with ease and swim expertly underwater. They are, however, nearly helpless on land. Loons are typically dagger-billed, with feet set back; the sexes look alike. They nest on inland lakes in the Far North and winter mostly on salt water along both the Atlantic and Pacific coasts. Normally a pair of loons raises two young in a mound nest, near or on the water, which is tended by both parents. Three species of loons are covered in this guide.

Grebes   Grebes (Podicipedidae) vary in size from those comparable to a small duck to some nearly the size of a small goose. They are expert swimmers and divers, and most have slender necks; straight, pointed bills; and feet set far back on the body. Like loons, they are nearly helpless on land. Grebes are water birds closely associated with inland lakes, ponds, and marshes, although some winter in coastal waters. Males and females are similar in appearance and raise two to ten young on a floating nest made of plant material. Six species of grebes are described in this guide.

Cormorants Cormorants (Phalacrocoracidae) are crow- to small-goose–sized water birds. Expert swimmers and divers, they subsist mainly on fish. When not in the water they often perch on pilings or rocks, displaying their webbed feet, an upright posture, and long, fish-catching bills. Cormorants often rest with wings outstretched in order to dry their feathers. They typically nest in small colonies, placing the nests on the ground or on a rocky ledge. Four species of cormorant are covered here.

Ducks, Geese, and Swans Ranging widely in size, from the smaller, pigeon-sized ducks to the massive swans, the ducks, geese, and swans in family Anatidae all have webbed feet and most have flattened and rounded bills. In general, these birds nest, feed, and/or roost in association with water. Many are highly migratory and thus seasonal in distribution. Males are generally larger than females. The sexes may be similar (as among geese and swans) or strikingly different (as among many ducks). Most of these birds are solitary nesters, typically choosing the edge of a wetland or a nearby area. Nests are often constructed of reeds and grasses and lined with down; a few species nest in tree cavities. Fifty-eight species of ducks, geese, and swans appear in this guide.

Gallinules, Coots and Moorhens
The gallinules, coots, and moorhens of family Rallidae are mostly chicken-sized water and marsh birds with short wings. Some species are highly secretive. The toes of these birds are often adapted for walking on aquatic vegetation and muddy flats. Mainly migratory and thus seasonal in distribution, these birds build platform-type nests that are typically well hidden in the emergent marsh- or pond-side vegetation. Three species of gallinules, coots, and moorhens are covered here.

THE BIRDS

Red-throated Loon *Gavia stellata*

Superbly adapted to an aquatic life, all loons breed on freshwater lakes or ponds and migrate to salt water in winter, as inland water freezes. Though all loons are superficially similar while in their winter plumages, more extensive white on the face of winter Red-throated Loons helps distinguish them from their relatives, the Pacific and Common loons (see pages 26 and 28). The slimmest of the loons, Red-throateds also have a finer bill.

Identification
24–27". Dark gray back is finely speckled with white; underparts are white. In breeding plumage, it shows a gray head and neck and a brick-red throat-patch. In winter, the face from the top of the eye down and the foreneck are white.

Voice
Seldom heard, except on breeding grounds, where it utters a variety of wails, shrieks, and a low growl.

Habitat
Tundra and coastal lakes and ponds in summer; coastal bays and estuaries in winter.

Range
Circumpolar arctic south to Manitoba and Newfoundland; winters along the Atlantic and Pacific coasts; casual on the Great Lakes.

24

Pacific Loon *Gavia pacifica*

The Pacific Loon and the Arctic Loon (*G. arctica*) were formerly considered to be one species, referred to as the Arctic Loon. Their diet on the wintering grounds is almost exclusively fish, while in summer they also eat insects, crustaceans, small amphibians, and even plants. Pacific Loons migrate during the day, and they normally travel in small flocks consisting of loose formations.

Identification 23–29". In breeding plumage (as shown here), the crown and nape are pale gray and contrast with the dark back and throat. The iridescent purple throat is noticeable only in good light. In winter this species is basically dark gray above and light below. Its relatively small size and slim, straight bill are characteristic.

Voice A yodel similar to that of the Common Loon, but higher and ascending in pitch. Also, a soft *kwow.*

Habitat For breeding, large freshwater lakes; in winter, coastal waters.

Range Breeds in Alaska and the Far North; winters along the Pacific Coast.

26

Common Loon *Gavia immer*

Loons are large water birds that normally winter along the coast and move inland during the breeding season. The Common Loon seems to prefer isolated lakes, particularly those with numerous bays and islands. Human activities, such as motor-boating, may deter Common Loons from nesting or even cause the parents to abandon their nest. Common Loons are excellent divers and swimmers and rely on a diet of fish.

Identification 28–36". A large black and white bird with a thick bill. In flight its long neck and head are stretched forward, and the legs and feet project behind. Common Loons often float low in the water, and regularly dive for fish.

Voice A loud, laughing yodel and a mournful wail, often heard at night on breeding territory.

Habitat Large, forested lakes with many small islands; oceans and bays in winter.

Range Breeds primarily in the North, from Alaska east through northern Canada and in northern forested states. Winters along both coasts and as far south as the Gulf of Mexico.

28

Pied-billed Grebe *Podilymbus podiceps*

Grebes are strictly water birds; they rarely set foot on land. A quick inspection of the grebe's anatomy reveals that its legs are set so far back on its body that they are essentially useless for walking on dry land. In its chosen habitat, however, the grebe's leg position and lobed toes move it efficiently on and under the surface of the water. The Pied-billed Grebe, one of the smaller North American grebe species, is named for its two-colored bill, which is apparent during the breeding season.

Identification 12–15". A chunky, plain-looking brown bird with a white rump, invariably seen on the water. It has a pied bill and a black throat in summer and a plain, pale bill and white throat in winter.

Voice Repeated calls include clucking and hollow cooing: *coo coo coo cuk cuk cuk cuk*; also a nasal whinny.

Habitat Freshwater ponds, marshes, and sluggish streams; salt water in winter if freshwater habitat is frozen.

Range Widespread in North America.

Horned Grebe *Podiceps auritus*

The Horned Grebe, like many other birds, leads a radically different life in summer than in winter. Its outward appearance and its chosen habitat vary greatly with the seasons. During the summer, this gaily feathered bird is strictly a resident of inland, freshwater wetlands. With the arrival of winter, however, the Horned Grebe's plumage takes on a muted character, and the bird abandons its freshwater nesting territories in favor of saltwater coasts.

Identification 12–15". A smallish, ducklike bird, invariably found around water. In summer, it shows a rufous foreneck and flanks, as well as golden buff ear tufts. In winter, it is black above and white below, with the white extending up onto the cheek.

Voice Loud shrieks, and chatters that sound like a tape recording played on fast-forward.

Habitat In summer, lakes and marshes with open water; in winter, mostly on salt water and also on the Great Lakes.

Range Breeds from Alaska and northern Canada south to the Great Lakes. Winters along coastal waters.

Red-necked Grebe *Podiceps grisegena*

While Red-necked Grebes migrate in loose flocks, breeding pairs are known to be very shy, often remaining out of sight on their large nesting territories. Their nest consists mainly of grasses, and it often floats near the edge of an inland lake. Red-necked Grebes sometimes feed their chicks feathers. It is believed that the feathers help strain out bones and other undigested matter, which is then regurgitated.

Identification   18–22". One of the larger North American grebes, this is the largest normally found on the East Coast during winter. In winter it is basically dark above and light below with a long yellow bill. A contrasting white chin strap forms a crescent on the side of the face. Breeding plumage shows a white cheek with a reddish throat and upper breast.

Voice   A variety of wails, staccato brays, and nasal grunts.

Habitat   Marshy ponds and lakes in summer; coastlines in winter.

Range   Breeds from Alaska across central Canada to Ontario, south to Washington, Montana, North Dakota, and Minnesota. Winters along the Pacific and Atlantic coasts.

Eared Grebe *Podiceps nigricollis*

This gregarious water bird nests in large colonies and collects in huge flocks at its wintering quarters. On the Salton Sea in southern California, over a million Eared Grebes congregate each year. When frightened, grebes often slowly sink into the water, leaving only their heads exposed.

Identification 12–14". In breeding plumage birds are dark brown and blackish, with rusty flanks and straw-colored ear tufts (for which they are named). The black head is crested. In winter plumage there is a limited amount of white on the upper throat, the chin, and behind the ear; the underparts are white. White patches in the wing are visible in flight.

Voice Usually silent, except during the breeding season, when it utters froglike peeps and squeaky whines.

Habitat Marshy freshwater lakes and ponds during the breeding season; open freshwater lakes and ocean bays in winter.

Range A Western bird. Breeds inland from British Columbia and Manitoba south to Texas. Winters along the Pacific and Gulf coasts; casual visitor to the Atlantic Coast.

36

Western Grebe *Aechmophorus occidentalis*

Two species of grebes that occur mainly in the western United States look so similar that until recently they were considered to be one species. The Western Grebe and Clark's Grebe (see page 40) both breed on western lakes and winter along the Pacific and Gulf coasts. They can be distinguished by their different facial patterns. Western Grebes are well known for their "dancing-on-water" courtship displays.

Identification	22–29". A large grebe, somewhat resembling a loon. It is overall black and white with a long, slender neck and a long, thin, yellowish-green bill. The black head markings extend below the eye, creating a masklike effect.
Voice	A two-noted *crick crick!* with the tone like that of a cricket but less measured.
Habitat	Large, freshwater lakes with reeds and rushes for nesting. Winters along coastal shores, bays, and large inland lakes.
Range	Breeds from central Canada south to New Mexico. Winters mainly along the Pacific and Gulf coasts and inland in southern New Mexico.

38

Clark's Grebe *Aechmophorus clarkii*

This species and the similar Western Grebe (see page 38) were considered to be one species until relatively recently. As is true of most grebes, both male and female Clark's Grebes take active parts in the nesting cycle. The adults share incubation chores, and soon after the downy young are hatched, the parents accompany them on their first excursions onto open water. Often the young are carried on a parent's back and ferried to and fro while it fishes.

Identification	22–29". A large grebe, somewhat resembling a loon. It is overall black and white with a long, slender neck and a long, thin, yellowish-orange bill. The black head markings extend onto the face, but the black cap does not reach the eye.
Voice	A single *crick!*, the tone like that of a cricket.
Habitat	Large, freshwater lakes with reeds and rushes for nesting. Winters along coastal shores, bays, and large inland lakes.
Range	Breeds from central Canada south to New Mexico and is more common in the western and southern portions of the range. Winters mainly along the Pacific and Gulf coasts and inland in southern New Mexico.

40

Great Cormorant *Phalacrocorax carbo*

This large blackish bird once was called the European Cormorant because it colonized the northeastern coast of North America from Europe. Its habit of sitting upright on conspicuous perches, drying its outstretched wings, is typical of most cormorants. But unlike the Double-crested Cormorant, the only other cormorant with which it shares a range, the Great Cormorant is found almost exclusively on the coast. The Double-crested migrates farther south.

Identification 36". A large bird with a heavy, hooked bill. Adults are entirely black with traces of iridescence. They show bare, orange skin around the chin and a whitish throat. In breeding plumage, a white patch on the lower flank is sometimes concealed by the folded wing. There is a variable amount of fine white streaking on the head.

Voice Deep, guttural grunts.

Habitat Rocky shores, coastal islands, and seaside cliffs.

Range In North America, breeds locally from northeastern Canada to Maine. Winters along the East Coast.

42

Double-crested Cormorant *Phalacrocorax auritus*

A cormorant is a large, dark seabird often seen perched on a piling or buoy with its wings outstretched. Cormorants spend much of their time fishing underwater, and this spread-winged pose allows their not-quite-waterproof feathers to dry. Although they have never been domesticated, various species of cormorant have assisted humans. The guano industry in South America is a direct beneficiary of the cormorant, and these birds have been used by fishermen to help catch fish.

Identification	30–36". A large, though slim, long-necked species with an orange throat patch. The thin, hooked bill and the upward tilt of the head are apparent when the bird is swimming. It often migrates in large numbers in gooselike formations.
Voice	Normally silent, except for grunting calls in the breeding colony.
Habitat	Lakes, rivers, and rocky coasts; bodies of water that provide plentiful fishing.
Range	Widespread in North America. Winters along the coasts, except in the Northeast.

44

Neotropic Cormorant *Phalacrocorax brasilianus*

The Neotropic was once called the Olivaceous Cormorant, in reference to the olive-green highlights on its plumage. These are often obscure, however, and the bird normally appears brown. This smallest North American cormorant, unlike other members of the family, may perch on wires. More typically, however, it perches with wings outstretched on dead snags along the edges of ponds or lakes. In Mexico, Neotropics have been observed hunting collectively, driving fish by beating the water with their wings.

Identification 26". A relatively small cormorant, overall glossy dark with a yellowish throat-patch that has a characteristic white, V-shaped border during the breeding season. Within its range it might be confused with the Double-crested Cormorant, but the Neotropic is smaller and has a longer tail.

Voice Generally silent.

Habitat A wide variety of salt- and freshwater habitats including marshes, seacoasts, rivers, and lakes.

Range A resident of coastal Texas and Louisiana, and scattered inland sites in the Southwest.

46

Pelagic Cormorant *Phalacrocorax pelagicus*

This is the smallest cormorant found on the West Coast. Its small head and slender neck and bill help distinguish it from its larger relatives. Found primarily along the most rugged and exposed stretches of coastline, it is at home fishing in turbulent surf. Being lighter-bodied than other cormorants, it does not struggle to get airborne, and occasionally it launches directly into the air upon surfacing from a dive.

Identification 25–30". A small and slim cormorant with a small head, slender neck, and dark, thin bill. The adult is glossy black with a small white patch on the lower flank. Immatures are uniformly dark brown.

Voice Groaning and hissing notes, heard around the breeding colonies.

Habitat Bays, seaside cliffs and islands, rocky shores, and coastal waters. Feeds coastally or at sea.

Range A year-round resident of the Pacific Ocean; seen along coastal North America from Alaska to Baja California.

48

Fulvous Whistling-Duck *Dendrocygna bicolor*

The two species of North American whistling-ducks, with their long necks and legs, look somewhat like small geese. Interestingly, this species was formerly called Fulvous Tree Duck, even though it is seldom found in trees. Fulvous Whistling-Ducks often travel in small groups, wandering widely in search of food. Their diet consists mainly of aquatic vegetable matter, and they are especially fond of wild grains and rice. Their nest, constructed of grasses and sedges, may be fitted with a domelike roof. Typically a dozen eggs will constitute the clutch, but one nest was found with approximately 100 eggs, apparently laid by several females.

Identification 18–21". A relatively large duck. The body is rich tan or fulvous, and the wings dark. The legs and bill are dark gray. In flight, the characteristic white tail-band may be seen.

Voice A thin, almost hoarse, whistle.

Habitat Marshlands, flooded farmlands, and rice fields.

Range Summers throughout the southern third of the U.S., sporadically from southern California east to Virginia. Resident in southernmost regions of Texas and Florida.

50

Black-bellied Whistling-Duck *Dendrocygna autumnalis*

This is the less common of the two species of whistling-ducks found in North America. It was formerly called the Black-bellied Tree Duck, for its habit of nesting in tree cavities. Females lay from 12 to 16 eggs, which are incubated for nearly a month. Soon after the chicks hatch, the female leaves the nest and calls to the young. The downy chicks then leap from their nest, following the parent to a nearby pond or marsh. Young birds may remain with the adult for up to six months after hatching.

Identification 21". A relatively large duck with a gray face and white eye-ring. The bill and legs are pinkish red. The lower neck and breast are rich tan, contrasting with the black belly. Birds in flight show a white area contrasting with the black trailing edge of the wing.

Voice In flight, flocks utter a series of squeals and whistles.

Habitat Tree-lined ponds and riverbanks, marshes and swamps.

Range Breeds in south-central Arizona and is a resident of extreme southern Texas. Casual visitor to southeastern California and Louisiana.

Tundra Swan *Cygnus columbianus*

Formerly called the Whistling Swan, in reference to Lewis and Clark's early description of its vocalization, this bird's name now refers to the species' arctic nesting areas. Tundra Swans are known for their wide-ranging flights. During the winter, small groups of Tundra Swans may be found among larger concentrations of Canada Geese.

Identification 47–58". The smallest swan species in the U.S., this bird is white with a black bill and black facial skin. Normally a contrasting yellow spot is apparent just in front of the eye. The neck is held straight and stiff.

Voice Similar to the honking of a Canada Goose, but higher in pitch, more melodious, and softer.

Habitat Nests in tundra island pools and marshes; winters on lakes, ponds, and coasts.

Range Breeds in northwestern Alaska along the coast and in the Arctic; winters coastally from Washington to California, inland from California to Nevada, and in Utah, Arizona, Texas, and on the East Coast from New Jersey to North Carolina.

54

Whooper Swan *Cygnus cygnus*

The Whooper Swan is a Eurasian species found only rarely in North America. Birdwatchers wishing to add this to their life lists will probably have to travel to the Aleutian Islands off the coast of Alaska. Known as the noisiest of the Eurasian swan species, the Whooper typically nests in tundra and coastal wetlands. While this species is closely related to North America's Trumpeter Swan, distinguishing between the two species is fairly easy.

Identification 60". Adult birds are largely white and are best identified by the large yellow patch covering the area between the eye and the nostril. The bill is black.

Voice A low-pitched *whoop-whoop.*

Habitat Nests in inland marshes and shallow tundra lakes and ponds; winters on seacoasts.

Range Winters in Alaska's Aleutian Islands.

Trumpeter Swan *Cygnus buccinator*

The Trumpeter Swan is the largest swan species in the world and the largest waterfowl in North America. During the 19th century this magnificent bird was trapped and hunted for its meat, skin, and feathers, and pressured nearly into extinction. Fortunately, strict legislation and reintroduction programs have helped. While still uncommon, Trumpeter Swans can be found at several National Parks and Wildlife Refuges, including Yellowstone.

Identification	60–72". Adult birds are mainly white; young birds are dusky gray. Adults have all-black bills; juveniles' bills are pinkish with a black base. This swan's neck is held with a bend at the base but is otherwise straight.
Voice	A series of trumpetlike, single or double honks on one pitch.
Habitat	Lakes and ponds; also coastal waters in winter.
Range	There are two breeding populations in North America. One migratory group breeds in Alaska, Alberta, and Saskatchewan, and winters along the coast from Alaska to Washington. The other group, mainly resident, is scattered across the western and northwestern U.S.

Mute Swan *Cygnus olor*

North America has two native swans—the Tundra and the Trumpeter Swan—but the Mute Swan is an introduced species. Mute Swans were brought to the United States from Europe more than a century ago. Like other introduced species, naturalized Mute Swans can cause problems for native species. Because they require, and are able to defend, large nesting territories, these swans often prevent native waterfowl from nesting in their traditional locales. The Mute Swan's aggressive behavior during the nesting season is not to be taken lightly; these massive birds have been known to attack and injure unsuspecting humans.

Identification 58–60". A very large, white bird with a wingspan to 8', an orange bill, and a black knob below the forehead. At rest, it shows a downward-tilted head and an S-shaped neck.

Voice Normally silent; occasionally hisses and grunts. Also rarely utters a loud trumpeting call.

Habitat Ponds, coastal lagoons, and open marshes.

Range From New England south to New Jersey, and around the Great Lakes.

60

Bean Goose *Anser fabalis*

The Bean Goose is a Eurasian waterfowl that occasionally strays to North America. Virtually all sightings occur in the Aleutian Islands and western Alaska. In Europe this species often feeds in agricultural fields; in fact, its common name is derived from its practice of visiting bean fields to forage. Its other colloquial names, including "harvest goose" and "cornfield goose," also refer to this species' favored feeding areas. On its summering grounds in northern Europe and Asia, the Bean Goose nests in small colonies, where a pair normally tend a clutch of four to five eggs.

Identification	31". A grayish-brown goose with orange feet and legs. The best distinguishing mark is the bill, which is black at the tip and base with an orange to pink splotch in the center.
Voice	Call is a low, reedy *ong-angk* on one pitch.
Habitat	A variety of habitats including northern tundra, alpine meadows, agricultural land, and marshes.
Range	Breeds in Eurasia and Greenland; a rare migrant in the central and western Aleutian Islands and the Bering Sea.

Pink-footed Goose *Anser brachyrhynchus*

While the Bean Goose (see page 62) is a vagrant to western North America, the Pink-footed Goose, a summer resident of Greenland, occasionally wanders to eastern North America. In fact, some authorities consider the Pink-footed Goose to be a subspecies of the Bean Goose. As in all sightings of exotic waterfowl, the observer should be aware that because many species of ducks and geese are kept by aviculturists, the possibility exists that any given bird is a domesticated individual.

Identification   26". A grayish-brown goose with pink feet and legs. The bill is black at the tip and base, with a pinkish splotch in the center.

Voice   A rather high-pitched, musical *ung-unk.*

Habitat   Tundra marshes and wetlands, agricultural areas, and coasts.

Range   Rare vagrant on the East Coast between Long Island, New York, and Newfoundland.

Greater White-fronted Goose *Anser albifrons*

Colloquially called "specklebelly" for the black markings on its brownish belly, the Greater White-fronted Goose occurs mainly in western North America. During the late spring and summer these geese nest on wet tundra in the high Arctic, feeding their chicks on swarms of aquatic insects. In late summer they move south in family groups. Winter congregations often number in the thousands.

Identification 26–34". An overall grayish-brown goose with a white band between the face and bill. The bill is normally pink (one race has an orange bill) and has a white tip. The brown belly is heavily marked with black barring and spotting.

Voice A high-pitched barking, or laughing sounds.

Habitat Higher inland tundra for breeding; open country, fields, and wetlands during migration and in winter.

Range In North America, breeds in Alaska and Arctic Canada; winters in the West along the Pacific Northwest coast, in California, and in parts of southwestern Montana, Idaho, Oregon, Nevada, New Mexico, the Texas panhandle, and coastally from Louisiana to Texas.

66

Snow Goose *Chen caerulescens*

This Arctic goose of North America has two distinct color phases. Birds from the western part of its range are largely white, while some of the eastern birds are largely dark. The dark-phase birds are referred to as "Blue Geese" and were formerly considered a separate species. The annual migrations of Snow Geese provide a great spectacle.

Identification 25–31". White-phase birds are overall white with distinctive black wingtips. Blue-phase birds have a white head and neck, contrasting with the largely brown body. Both phases have a pink bill and a characteristic black lip line.

Voice Shrill, doglike barking.

Habitat Nests in tundra grasses and marshes close to the sea; seen on migration and in winter feeding in fields and marshes.

Range Breeds in northern coastal Alaska, Arctic Canada, and along the western and southwestern coasts of Hudson Bay to James Bay. Winters in scattered locations: extreme southwestern British Columbia; in the southwestern U.S.; along the Gulf Coast of Texas and Louisiana; and along the mid-Atlantic Coast.

Ross' Goose *Chen rossii*

Closely related to the Snow Goose (see page 68), the smaller Ross' Goose shares some of its nesting areas, and the two species sometimes hybridize. Total numbers of Ross' Geese are relatively small, and its breeding range is restricted to the central Canadian Arctic. Like other geese, breeding adult Ross' typically forms close pair bonds. While the female does all the incubating, the male acts as a nest guard. Both parents assist the goslings in their early weeks.

Identification 21–25½". Smaller than the similar Snow Goose. Most are white with black wing tips; rarely a "blue"-phase bird occurs. The bill is pinkish and has a black mark at the base; the bill lacks the black lip line found in the Snow Goose.

Voice Staccato, nasal yelps and laughter: *anh anh, anh anh, anh anh anh.*

Habitat Coastal tundra in summer; grasses and grainfields on migration and in winter.

Range Breeds in the high Arctic; winters in southern Oregon, central and southern California, New Mexico, and extreme southern Texas.

70

Emperor Goose *Chen canagica*

The various geese species of North America can be separated into two groups: those that typically feed in fields, grazing on a variety of grasses; and those that are normally found on the coasts, foraging on saltwater or brackish vegetation and associated organisms. The Emperor Goose and the Brant (see page 74) are sea geese. The Emperor Goose is largely restricted to Alaska and the Aleutian Island chain. It nests at coastal ponds and marshes, or even in the driftwood along beaches. The Emperor Goose is regularly found foraging along the rocky seashore for seaweeds and a variety of shellfish.

Identification   26–28". A relatively small, stocky goose. Its white head contrasts with a silver-gray body that shows much black barring. The throat and foreneck are black. The bill is pinkish; the feet and legs are orange.

Voice   In flight, a hoarse *cla-ha, cla-ha.*

Habitat   Tidal ponds and marshes; rocky seashores and beaches.

Range   Breeds on the western coast of Alaska; winters mainly in the Aleutian Islands.

Brant *Branta bernicla*

The Brant is a sea goose, only rarely observed inland. While it breeds in the high Arctic, the Brant spends its winters along the Pacific and Atlantic coasts, foraging on a variety of aquatic vegetation. One important food source in winter is eelgrass. Beginning in the 1930s, eelgrass disappeared from many areas along the East Coast, and Brant were virtually extirpated from many areas.

Identification 23–26". Two forms occur: western birds, formerly referred to as Black Brant, have dark bellies; eastern birds have pale bellies that contrast more sharply with their upper parts. Both forms have a characteristic black head and neck with a contrasting white "collar." The upperparts are largely dark. In flight, the extensive white tail coverts are apparent.

Voice A low, guttural *crronk*.

Habitat The high Arctic tundra for nesting; in winter, coastal bays and eelgrass communities of the East and West coasts.

Range Breeds in the highest latitudes of the Arctic; winters along the Pacific Coast from British Columbia to California, and along the East Coast from Nova Scotia to Georgia.

74

Barnacle Goose *Branta leucopsis*

The Barnacle Goose is a vagrant to North America. It summers in Europe and Greenland, sometimes wandering to the Maritime Provinces and the northern Atlantic coast. Its name derives from medieval folklore, which held that migrant flocks appearing over the ocean had actually hatched from barnacles. The Barnacle Goose combines the habits of field geese and sea geese in that it often forages along the seashore but also flies inland to feed in fields and pastures. Because this bird is often kept by waterfowl enthusiasts, some North American sightings are undoubtedly of released or escaped birds.

Identification 27". This species has a characteristic white face and black neck-sock. There is a distinctive dark line running from the bill to the eye. The upperparts are bluish gray with black barring; the underparts are white.

Voice A doglike barking, similar to that of a terrier.

Habitat Arctic rivers and marshes; inland pastures and farms.

Range Breeds in Greenland and northern Eurasia; rare vagrant in the Northeast and inland to the Great Plains.

Canada Goose *Branta canadensis*

The Canada Goose is a common sight throughout much of the United States. On migration, V-shaped strings of these geese often call attention to themselves with their clamorous honking. Canada Geese are also seen roosting on ponds and lakes or grazing in fields. Seemingly comfortable around humankind, they regularly frequent golf courses, public parks, and neighborhood ponds. Because of their size and relative tameness, these geese are ideal subjects for making observations on bird behavior, especially during the nesting cycle. While one Canada Goose looks more or less like another, there is considerable size variation in the different races occupying North America.

Identification 22–45". A large waterfowl with a distinctive black neck and head and a white chin strap. Its body feathers are dusky below and dark brown above.

Voice A loud, two-noted honk, with second note higher in pitch.

Habitat Ponds, lakes, and other open waters; also feeds in fields, on golf courses, and in grasslands.

Range Widespread in North America.

78

Muscovy Duck *Cairina moschata*

This duck is a resident mainly of Mexico and Central and South America, but a few wild birds have been seen recently in the extreme southwestern United States. Observers spotting a Muscovy Duck, however, should consider the possibility that it is a domesticated Muscovy, as these birds are commonly raised by farmers and aviculturists. In the wild, male Muscovy Ducks vie with each other to become the dominant male in a flock.

Identification 25–35". A large, chunky species varying from black and white to all white. Males are significantly larger than females. Dark birds show glossy, green to purple upperparts and a white wing-patch. The area between the bill and the eye is often bare with knobby, red or black, wartlike protuberances.

Voice Generally silent, with some hissing from males; females quack.

Habitat Wooded streams, marshes, and swamps.

Range From Mexico to South America; in North America, recent sightings have occurred along the lower Rio Grande.

Wood Duck *Aix sponsa*

The male Wood Duck in full breeding plumage is an extraordinary sight. The patterning and colors on his head alone, a combination of glossy greens and bright reds, are as spectacular as those on any waterfowl in the world. While the female is definitely less decorative, her subtle hues of brown and blue are also a visual delight. As its name implies, this duck regularly frequents forested areas, and its nesting sites include cavities in mature or dead trees.

Identification 17–20". Males have glossy green head-feathers complete with a manelike crest. The eyes and base of the bill are bright red. Females are largely mouse-brown; however, there is a distinctive, tear-shaped, white eye-ring and a blue speculum.

Voice In flight, a high pitched *wooo-eeek*. Also soft mews and peeps. The female's whining flight call is frequently heard.

Habitat Wooded swamps, rivers, and ponds. In fall, freshwater marshes.

Range From Manitoba east to Nova Scotia and south to Texas. Also the Pacific Northwest east to Montana, and California.

Green-winged Teal *Anas crecca*

Of the three common teals of North America (the others are the Blue-winged Teal and the Cinnamon Teal), this is the smallest. On the wing, the Green-winged Teal is a delight to watch. Its rapid flight is regularly punctuated by sharp turns and twists. Although most Green-winged Teals winter in the southern United States and Mexico, they are an especially hardy species. Individuals regularly linger in the northerly parts of their range into early winter. In spring, Green-winged Teals are among the earliest migrants.

Identification 12–16". A small, dark duck. Males are overall grayish brown with a distinctive, chestnut-colored head and a contrasting swath of green from the eye to the nape. Females are largely buff-colored. In flight, the green speculum is seen.

Voice Males emit short, high-pitched whistles; females quack.

Habitat Marshy ponds, lakes, and rivers.

Range Breeds primarily from Alaska throughout Canada and in the western U.S. Winters throughout the southern half of the U.S.

Baikal Teal *Anas formosa*

Named for Lake Baikal in Siberia, this Asian species occasionally wanders to northwestern North America. Because this is another waterfowl species raised by aviculturists, many North American observations are probably of escapees.

Identification 17". Males have a distinctive head pattern, with a dark crown, and green and yellowish face-patches bordered in black. The male has long, buff-colored tertial feathers contrasting with his gray sides. Females closely resemble female Green-winged and Blue-winged teal. There is, however, a dark-bordered, pale, roundish patch at the base of the bill.

Voice Males utter a deep, chuckling *wot wot wot;* females quack.

Habitat In summer, swampy taiga, marshes, tundra river deltas; in winter, a variety of freshwater and brackish habitats.

Range Primarily Asia, but wanders to the Aleutian Islands, coastal Alaska, and as far south as coastal California.

Falcated Teal *Anas falcata*

The Falcated Teal is primarily an Asian species that is only rarely found in North America. It may be a relative of the Gadwall (see page 110) and is named for its sickle-shaped (falcated) tertial feathers. On its Siberian home range, this teal nests in streamside marshes and wet meadows in mountainous terrain. After the breeding season Falcated Teals move to coastal areas, where they pass the winter.

Identification 19". A chunky duck with a relatively large head. The male in breeding plumage has a distinctive, iridescent bronze-and-green head with a white throat. There is a pale spot at the base of the upper mandible. The body is overall gray with much black spotting and edging. Females look similar to the female American Wigeon.

Voice Males give a low-pitched whistle; females quack.

Habitat In summer, in marshes, swamps, lakes, and mountain and river valleys; in winter, also on salt water.

Range In North America, may be seen in spring and fall in the Aleutian and Pribilof islands.

American Black Duck *Anas rubripes*

The American Black Duck (seen here on the right) is declining in numbers, and the species is being pressured from several different directions. A reduction in the size of its wetland nesting habitat is a primary threat. The health of individual ducks is indirectly affected by the use of lead shot and pesticides, both of which introduce toxic substances into feeding areas. The species is also threatened by the Mallard: Because the two species readily interbreed, hybrids are common, and the number of "pure" American Black Ducks is declining.

Identification 19–22". Overall dark brown with a paler face and neck. The male's bill is yellow; the female's, greenish yellow. There is a purple speculum. In flight, seen from below, this duck appears dark, or blackish, with silvery wing linings.

Voice The quintessential duck quack.

Habitat Fresh- and saltwater marshes, ponds, swamps, and lakes.

Range An eastern bird. Breeds from Saskatchewan east to Prince Edward Island, south to New York. Winters south of its breeding range to the Gulf Coast.

Mottled Duck *Anas fulvigula*

The Mottled Duck is a close relative of the Mallard (see page 94). In fact, some waterfowl experts consider it a subspecies. Its behavior is somewhat different, however, in that it normally begins the breeding cycle in January, earlier than does the Mallard. Mottled Ducks may have laid their eggs by February, and the first downy young are seen less than a month after incubation begins. Like other duck species, Mottled Ducks may hatch six or more young, of which only one or two will survive their first few months. A wide range of predators, including alligators, snapping turtles, and garfish, cull each year's production.

Identification 22". Similar to the American Black Duck and the female Mallard (see pages 90 and 94). Distinguishing characteristics include an unmarked yellow bill and an unstreaked throat. The speculum is greenish blue with a black border.

Voice A loud quack.

Habitat Freshwater and coastal marshes, ponds, and farmlands.

Range Resident in Florida and along the Gulf Coast.

Mallard *Anas platyrhynchos*

Ducks can be divided into two basic groups, the dabblers and the divers. Dabblers feed from the surface, using tilting motions that allow them to reach aquatic vegetation, seeds, and other food just below the surface. The Mallard is the most abundant dabbler in North America and across the Northern Hemisphere. The progenitor of many duck species, the Mallard is capable of hybridizing with a wide variety of waterfowl, at times to the detriment of the other species. At home most anywhere, the Mallard takes readily to public places.

Identification   18–27". The male's green head and yellow bill are distinctive. Males also have a chestnut-colored breast and a blue speculum bordered in white. Female Mallards are overall light brown and also have a blue speculum bordered in white.

Voice   Females utter a loud, descending series of quacks; males utter softer notes.

Habitat   Commonly found on any shallow body of fresh water.

Range   Widespread in North America.

94

Spot-billed Duck *Anas poecilorhyncha*

Primarily an Asian duck, the Spot-billed occurs as a vagrant in the Aleutians and other Alaskan islands. Ornithologists recognize three subspecies of Spot-billed Duck; the Chinese Spot-billed is the most likely to occur in North America. Interestingly, this subspecies does not show the red spots on the bill for which the species was named.

Identification 24". Similar to the American Black Duck and the female Mallard (see pages 90 and 94). The head and foreparts are light tan, while the rest of the body is brown with a bluish speculum. The bill shows a distinctive pattern: largely black with a prominent yellow tip.

Voice Quacks.

Habitat Shallow, freshwater marshes and lakes with dense vegetation; may also be found on rivers.

Range An Asian species; vagrant to the Aleutians and other Alaskan islands.

White-cheeked Pintail *Anas bahamensis*

This species, also known as the Bahama Duck, is a resident in the Bahamas, the Virgin Islands, Hispaniola, and Puerto Rico. Individuals have been recorded in Florida and along the Gulf Coast. On its home range, the White-cheeked Pintail spends time on both coastal flats and inland ponds and feeds on aquatic plants and seeds. It has been known to hybridize with several North American resident ducks, including the Northern Pintail.

Identification 17–20". An overall brown duck with a white cheek and throat that contrast with a dark forehead, nape, and hindneck. The distinctive bill is largely blue with a prominent red mark near the base. The long tail and tail coverts are buff-colored. In flight, the green speculum bordered with tan can be seen.

Voice Usually silent. The males may whistle; the females give a descending series of quacks.

Habitat Largely coastal waters, but also inland salt- and freshwater ponds and lagoons.

Range May be seen in North America as a casual vagrant to southern Florida and elsewhere along the Gulf Coast.

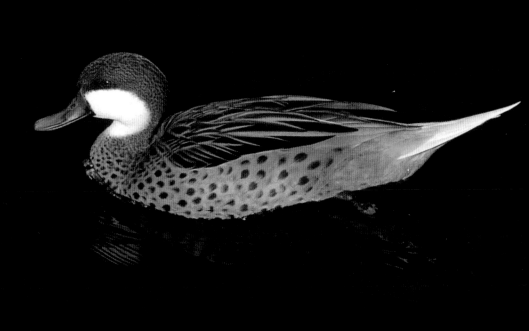

Northern Pintail *Anas acuta*

The Northern Pintail, one of North America's more abundant duck species, forms pairs during the fall and winter. The female determines exactly which breeding ground they will migrate to in spring. The Northern Pintail's nest is often located a considerable distance from water. As many as a dozen eggs may be laid.

Identification   25–29". A relatively large duck, but slender in overall appearance. The male's brown head and white neck are distinctive. His body plumage is largely gray, and he has a characteristically long, narrow tail. Females are overall light brown to tan. Pintails have bluish bills.

Voice   Males give a clear, two-noted whistle; also a nasal *unk unk unk.* Females quack.

Habitat   Marshes, wet tundra, and shallow lakes; also along the coast in winter.

Range   Breeds primarily in the North, from Alaska east to Prince Edward Island, and also in the western U.S. as far south as New Mexico. Winters on the West Coast south to Texas and along the East Coast and the Great Lakes.

Garganey *Anas querquedula*

This Eurasian species is still rare in most areas of North America, but it has shown up here with increasing frequency over the last decade. Most sightings are from Alaskan waters and south along the West Coast to California, but there are several records from as far east as the Caribbean. This species is most likely to be confused with other teal (it is sometimes called the Garganey Teal), especially the Blue-winged Teal (see page 104).

Identification   15–16". In breeding plumage, males are readily identified by the broad white stripe beginning over the eye and ending on the side of the neck. The crown is dark and the face is brownish red. Females are most similar to female Blue-winged Teal, but they have a more patterned face.

Voice   Males utter cackling sounds; females quack.

Habitat   Shallow pools and marshes with plenty of vegetation for cover and nesting; rarely on salt water.

Range   May be seen during migration in the western Aleutian Islands; rarely on the East and West coasts or in the Midwest.

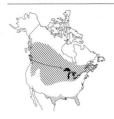

Blue-winged Teal *Anas discors*

Blue-winged Teal on the wing, with their rapid flight and acrobatic maneuvers, are a joy to behold. These teal seem to be somewhat less hardy than other duck species. They are often late arrivals in spring and are among the first ducks to move south in fall. Blue-winged Teal may be found throughout much of the United States and Canada during the breeding season. They nest in the grasses and reeds on or near the edges of marshes and prairie "potholes."

Identification 14½–16". A relatively small duck. The male is distinctive, with his blue-gray head sporting a prominent white crescent. Females are mottled buff-brown and similar to female Cinnamon Teal (see page 106). In flight, the large powder-blue wing-patch and green speculum can be seen.

Voice A squeaky *peep, peep, peep;* also a soft, barking quack.

Habitat Shallow, freshwater marshes and ponds.

Range Breeds from southeastern Alaska to the East Coast, and southward across much of the United States. Winters from North Carolina south to Florida and along the Gulf Coast to Texas.

104

Cinnamon Teal *Anas cyanoptera*

Identifying female and juvenile ducks is often difficult. A good case in point are the three common teal species covered in this guide. All are small, rather plainly marked, brown ducks. Advanced identification guides refer to subtle differences that are often relative in character. Only experience in the field will make such tips useful. However, range information may be a useful starting point in the identification process.

Identification
14–17". Drakes are characteristically rufous below and on the neck and head. Both sexes display a blue wing-patch. Female and immature birds are largely buff-brown. The female has darker plumage, less distinct facial markings, and a larger bill than the female Blue-winged Teal.

Voice
Males utter a low chatter; females, a soft quacking.

Habitat
Shallow lakes, ponds, prairie marshes, and sluggish streams bordered by grasses and reeds.

Range
A western bird; breeds from southern British Columbia south to Texas; year-round in the southwestern states.

Northern Shoveler *Anas clypeata*

The Northern Shoveler is closely related to the Blue-winged Teal. Like other surface-feeding ducks, called dabblers, the shoveler uses comblike structures on the edges of its bill as a strainer. It skims a variety of small plants and animals off the water's surface, but it also takes food from the bottoms of ponds or lakes. Northern Shovelers often nest near prairie wetlands or ponds. Within hours of their hatching, the chicks are led by the female to a nearby marsh.

Identification 17–20". The large, spatulate bill and flattish head of this duck are distinctive. Males have a green head, white breast, and conspicuous chestnut sides. Females are drab brown. Both sexes show a blue wing-patch and a green speculum.

Voice A low croak, chuckle, or quack.

Habitat Shallow lakes and marshes.

Range Breeds from Alaska south and east to Saskatchewan and Manitoba, south to California, New Mexico, and east to the Great Lakes Region; also occasionally throughout the Northeast. Winters from Oregon and California south to Texas and along the Atlantic Coast north to New Jersey.

108

Gadwall *Anas strepera*

While the Gadwall is a dabbler, getting much of its food in the typical head-down, tail-up position, this species is also an accomplished diver, collecting roots and tubers from pond and lake bottoms. Drake Gadwalls, like other male ducks, take a limited role in family matters. Soon after pairing and copulation, males leave the female and join other males. The hens take on the responsibilities of nest-building and incubation, and they shepherd the downy young on their first trip to the water.

Identification 18½–23". Males are mostly gray; females, mostly mottled brown. Both sexes have white bellies and specula. The male has a distinctive black rump.

Voice Males utter *kack kack* sounds and whistles; females quack.

Habitat Freshwater marshes, ponds, and rivers; also salt marshes in some locations.

Range Breeds from southern Alaska, western British Columbia, and south-central Canada to the Great Lakes, and throughout the western and midwestern U.S. Winters across the southern half of the U.S.

110

Eurasian Wigeon *Anas penelope*

This Eurasian species, closely related to the American Wigeon, occurs in North America as an uncommon vagrant along both the Atlantic and Pacific coasts. Most records for North America occur during the winter and early spring. On its home range the Eurasian Wigeon (sometimes referred to as the European Wigeon) nests on tundra or moorland grasslands.

Identification 16½–20". Similar to the American Wigeon (see page 114). The male's head is largely fox-red with a broad, cream-colored forehead and crown. The dark head and neck contrast with the lighter gray body feathers. Females are much like American Wigeon females.

Voice Males emit a clear, two-noted whistle; females quack.

Habitat In summer, small lakes and marshes in open or partly wooded country; in winter, flooded grasslands and wet meadows, coastal waters, including estuaries, and lakes.

Range Regularly vagrant along the Atlantic and Pacific coasts and in the Aleutian Islands in fall and winter; casual inland during spring.

112

American Wigeon *Anas americana*

The American Wigeon obtains most of its plant food from the shallow edges of ponds and lakes. But one favored food is wild celery, which grows in deeper water. The American Wigeon must rely on the efforts of diving species to tear the plant loose from the bottom; then the wigeons partake of the feast. They may even pluck a morsel from the bill of a passing coot.

Identification   18–23". The colloquial name "bald pate" derives from the male's whitish crown and forehead. Males also sport a green face-patch sweeping back from the eye. Females have a grayish head and a mottled brown body. In flight, the white belly and white patch on the upperwing are apparent.

Voice   Males give a series of soft whistles; females, a low *quawk, quawk, quawk.*

Habitat   Freshwater ponds and marshes, prairies, and tundra; also salt marshes in winter.

Range   Breeds from Alaska to southern Quebec, south to Nevada, Colorado, the Dakotas, and Minnesota; winters along all three coasts, Florida, and the extreme Southwest.

114

Common Pochard *Aythya ferina*

Extreme northwestern North America appears to be on the edge of the migratory route of this species. Nesting primarily in Eurasia, it is found with some regularity in the Pribilofs and the Aleutian Islands.

Identification 18–22". Similar to the Redhead (see page 120) and the Canvasback (see page 118). Males have a red head, black chest and rump, and grayish body. Females are largely tawny brown. The black tip of the bill is similar to that of the Redhead. However, both males and females have a distinctive blue-gray band crossing the mid-bill area.

Voice Males give a soft, nasal wheeze; females, a harsh *karr.*

Habitat Breeds on inland waters of steppes and plains; winters on large interior lakes and coastal bays.

Range Nests in Eurasia; spring migrant and summer visitor to the Pribilofs and the Aleutian Islands.

Canvasback *Aythya valisineria*

The Canvasback is one of the largest diving ducks in North America. It is prized by many hunters, a fact that may account partially for its reputation of being very wary. This and other waterfowl that bottom-feed are increasingly being affected by lead poisoning from spent shotgun pellets. The use of steel shot is now being advocated.

Identification	19½–24". Males have a rusty-red head, black chest and rump, and grayish-white body. Females have a pale brown head and neck and a grayish-brown body. There is a characteristic shallow, sloping forehead and a plain black bill. (See also the similar Redhead, page 120.)
Voice	Generally silent except in courtship, during which males utter a soft *coo* and females, a harsh *krrr,* or purring sound.
Habitat	Prairie ponds and lakes with border vegetation; winters on open fresh and salt waters.
Range	Breeds from Alaska to Minnesota and is resident in the northwestern U.S. to Colorado. Winters on the Pacific Coast from Washington to California, throughout the southern half of the U.S., and in the eastern U.S.

Redhead *Aythya americana*

Redheads make their summer home on the prairies of the United States and Canada. Because of the large-scale draining of wetlands in the East, nesting Redheads there are now much reduced. The female Redhead may lay seven to nine eggs and incubate them herself. Or she may incubate one set of eggs and deposit several other eggs into another female's nest (this practice is called "dumping").

Identification 18–22". Males have a rusty-red head, black chest and rump, and grayish-white body. Females have a pale brown head and neck and a grayish-brown body. The rounded head and the bluish bill with its black tip are distinctive.

Voice In courtship, males give a *mew mew mew;* females quack.

Habitat Prairie lakes and marshes in summer; protected coastal waters, marshes, and lakes in winter.

Range Breeds in south-central Alaska, British Columbia east to Minnesota, and east of the Great Lakes to northern New England; throughout the western U.S. as far south as Colorado. Winters in the southern half of the U.S. and along the Atlantic Coast as far north as Massachusetts.

Ring-necked Duck *Aythya collaris*

Some birds seem to be named for their least obvious field mark. Such is the case with the Ring-necked Duck. Males do have a chestnut-colored ring around their necks, but it is seldom visible in the field. Ring-necked Ducks are among the early migrants of spring, arriving at their breeding grounds by early May. If the winter is not particularly severe, they will often remain quite far north into January.

Identification 14½–18". Males have a black head, chest, and back with gray flanks. A white vertical stripe is often visible on the side. Females are largely brown. Both sexes have slightly pointed heads and black-tipped, grayish bills ringed with white.

Voice Usually silent, except during courtship, when males give a low-pitched whistle; females growl.

Habitat Freshwater lakes, ponds, and marshes in open country in summer; winters on lakes and protected coastal waters.

Range Breeds from Alaska across Canada to Nova Scotia and south to California, South Dakota, the Great Lakes, and northern New England. Winters on the Atlantic and Pacific U.S. coasts, and in the southern half of the U.S.

122

Tufted Duck *Aythya fuligula*

The Tufted Duck is a close relative of the scaups and similar in appearance to the Ring-necked Duck (see page 122). It is primarily a Eurasian species but shows up on both North American coasts with some regularity. While somewhat regular in western Alaska, the Tufted Duck normally makes its rare visits southward during the winter months. Some waterfowl experts feel that it will eventually become a breeding species in North America.

Identification   17". Males have a black head, chest, and back with white sides. Females are largely brown with buff-colored flanks. The head is more rounded than in the Ring-necked Duck, and a drooping crown tuft is often present.

Voice   Generally silent, except during courtship, when the male whistles and the female growls.

Habitat   A variety of lowland freshwater lakes, ponds, and rivers with border vegetation; winters also on sheltered coastal waters.

Range   May be seen in the U.S. in western Alaska, and in winter along the East and West coasts.

Greater Scaup *Aythya marila*

Known colloquially as the "broad-bill" or "blue-bill," the Greater Scaup, and its slightly smaller relative the Lesser Scaup (see page 128), present one of the more difficult problems of identification. Away from their breeding grounds, scaup are usually found in large flocks on water. The Greater Scaup shows a preference for salt water.

Identification 15½–20". The male has a finely barred gray back, black breast, bright white sides, and an iridescent blackish-green head. Its bill is pale blue. The female is dark brown with white around the base of the bill. This bird is distinguished from the Lesser Scaup by a dark greenish iridescence on a rounder head, and a longer white wing stripe, seen in flight.

Voice Common call a loud *scaup;* females give a low *arr.*

Habitat Lakes, bays, rivers, and tundra ponds. Winters on seacoasts and inland on large lakes.

Range Breeds in Alaska and northern Canada south to Quebec; irregular summer resident south to Michigan and Nova Scotia. Winters on the East and West coasts, the Great Lakes, and less commonly in the Mississippi Valley.

Lesser Scaup *Aythya affinis*

The Lesser Scaup is more often associated with fresh water than its larger relative the Greater Scaup (see page 126). Also called "broad-bill" or "blue-bill," the Lesser Scaup is an expert swimmer and diver. Both species of scaup feed on aquatic plants as well as insects and crustaceans.

Identification
15–18½". Males have a finely barred gray back, black breast, grayish-white sides, and an iridescent purplish head. The bill is pale blue. Females are dark brown with white around the base of the bill. This bird is distinguished from the Greater Scaup by a purplish iridescence on the head and a shorter white wing stripe, seen in flight.

Voice
Common call a loud *scaup;* females give a drawn-out, purring *kwuh.*

Habitat
Marshes, tundra ponds, and lakes. Winters on seacoasts and inland on lakes and reservoirs.

Range
Breeds in Alaska and northern Canada through Manitoba and south through northeastern Colorado and Iowa. Winters along the East and West coasts as well as inland south of Colorado and the Great Lakes to the Gulf Coast.

128

Common Eider *Somateria mollissima*

This largest of sea ducks is best known as a valuable source of insulation for sleeping bags and parkas. Though all birds have down feathers, duck down is particularly thick, and eider down is the most efficient insulator of all. In winter, especially in New England waters, flocks of Common Eiders form immense feeding rafts on the ocean above shellfish beds. Eiders typically migrate in long, undulating lines, low over the water.

Identification 23–27". A heavy-bodied duck with a relatively long, heavy bill that imparts a "Roman nose" profile. Adult males have a white back and breast and black sides, tail, and crown. The rest of the head is white with a greenish tinge visible at close range. Females are uniformly dark brown, varying in tone. Immature males are variably black and white.

Voice Courting males utter a pigeonlike cooing or moaning; females, a grating *gog-gog-gog*.

Habitat Seacoasts.

Range Breeds along Arctic coasts; winters along the northern Atlantic and northern Pacific coasts.

130

King Eider *Somateria spectabilis*

Eiders are northern sea ducks (see Common Eider, page 130). The King Eider summers in the Arctic and is one of the common nesting ducks of the tundra. Shellfish and crustaceans are its favorite food, and it is an expert swimmer and diver. In late summer large flocks of eiders gather in open bays in preparation for migration. Male King Eiders normally precede the females and young southward.

Identification 18½–25". Males are generally white on the head, neck, and breast and black on the back and flanks. They sport a large orange knob, or forehead shield, above the bill, and a light gray cap. Females are overall buff-colored.

Voice Males give a cooing moan in courtship; females growl or croak.

Habitat Coastal and inland tundra in summer; open coastal waters in winter.

Range Breeds in the Arctic; may be seen in winter in the Aleutians and rarely as far south as California, and off the Atlantic Coast from Newfoundland to Long Island, New York.

Spectacled Eider *Somateria fischeri*

The Spectacled Eider is a northern sea duck closely associated with Alaskan waters. In contrast to the Common Eider, which occurs in vast flocks, this bird has never been recorded in large numbers, and it is apparently relatively uncommon throughout its range. Spectacled Eiders combine the attributes of dabbling ducks and diving ducks in that they feed on insect larvae and plants on their nesting ponds while diving for mollusks in coastal waters. Interestingly, their whereabouts during the winter are still unknown.

Identification 20½–22". Drakes have a distinctive pale green head, yellow bill, and large, spectacle-like white eye-patch rimmed in black. The male's body is largely white above and black below. Females are mottled brown with a light buff spectacle. In both sexes the large bill is partially feathered.

Voice Generally silent (or rarely heard), except for some crooning and cooing sounds.

Habitat Coastal tundra pools and rivers in summer; probably winters at sea.

Range Breeds on northern and western Alaskan coast.

Steller's Eider *Polysticta stelleri*

Steller's Eider was named for the 18th-century German zoologist and explorer Georg W. Steller. This is the smallest and swiftest-flying of the eiders. Steller's Eider has a varied diet, including crowberries—a common berry of the Arctic tundra. The male, in a manner somewhat unusual for ducks, stays with the hen until incubation has begun.

Identification
17–18½". Males are largely black and white above and pale tan below with a black eye patch, a green-and-black nape crest, and a black collar. Females are mottled dark brown. Both sexes have a blue speculum bordered with white.

Voice
Barking, growls, and whistles near breeding grounds; otherwise, generally silent.

Habitat
Inland and coastal tundra pools in summer; winters on coastal waters and freshwater inlets.

Range
Breeds on the eastern Arctic coast of Alaska and south to St. Lawrence Island and the Yukon-Kuskokwim deltas; winters along the southwestern Alaskan coast and the Aleutian Islands, and casually south to the northern California coast.

136

Harlequin Duck *Histrionicus histrionicus*

Many birdwatchers consider this unique little duck to be one of the most beautiful species of ducks in North America. Exquisitely attired in a bold combination of steel-gray, black, white, and rich chestnut, these hardy birds spend their winters swimming and diving effortlessly among the rocks along exposed, surf-pounded shores.

Identification   14½–21". Adult males are unmistakable: steel-gray with bold slashes of black and white, a chestnut stripe over the eye, and a broad swath of chestnut along the flank. Females and immature males are dark brown with two white spots on the side of the face. They are smaller overall than female scoters, with a much smaller bill. Also compare with the female Bufflehead (see page 152).

Voice   Males emit high, squealing sounds; females, a harsh croak.

Habitat   Turbulent streams and rivers in summer; exposed rocky seashores in winter.

Range   Breeds in the West, from Alaska south to Wyoming, and in the East from the Arctic to northern Quebec. Winters along the northern Pacific and the northern Atlantic coasts.

Oldsquaw *Clangula hyemalis*

This sleek sea duck is not only one of the fastest of all ducks in flight, it is also one of the deepest divers. While foraging for shellfish, it may dive to at least 200 feet. The far-carrying, distinctive call for which it is named apparently reminded early American ornithologists of the voices of women. Its British name, Long-tailed Duck, seems more appropriate.

Identification 19–22½". The male is a combination of brown, black, and white—whiter in winter and darker brown in summer. Other than the male Northern Pintail, this is the only duck with very long, narrow tail feathers. The female shows variable combinations of dark brown and white; the light head with a dark cheek-patch is distinctive. Her tail is shorter than the male's.

Voice A yodel-like call, often heard in chorus.

Habitat Tundra ponds and marshes in summer; coastal shoals in winter.

Range Breeds throughout the Arctic. Winters in the coastal waters of the North Atlantic and North Pacific.

140

Black Scoter *Melanitta nigra*

The male Black Scoter is the only all-black duck in North America. A bright yellow-orange, fleshy protuberance at the base of the drake's bill gives it the colloquial names "butter-bill," "yellow-bill," and "yellow-nose," used by hunters. The Black Scoter is slightly smaller and lighter-bodied than the other two scoter species, and it takes flight from the surface of the water with less effort.

Identification 17–20½". The male is entirely black with a bright yellow-orange knob at the base of its black bill. The female is dark brown with a pale cheek-patch and light belly. The immature male is like the female but with a poorly developed, pale yellow knob at the base of the bill. In all plumages the undersurface of the flight feathers is silvery.

Voice The male may utter a whistle in courtship: *cour-loo*.

Habitat Tundra lakes and rivers in summer; seacoasts in winter.

Range In North America, breeds in western Alaska and northeastern Canada. Winters on both coasts.

Surf Scoter *Melanitta perspicillata*

This attractive sea duck is known by the less attractive colloquial name of "skunk-head" for its distinctive black-and-white head pattern. Due to the shape of the primaries, the wings of Surf Scoters produce a whistling sound in flight.

Identification 17–21". The male is black with white patches on the forehead and nape. It has a heavy, triangular bill that is bright red, orange, and white; the male also has white eyes. The female is dark brown with a pale belly and two whitish patches on the side of the head. The immature male is like the female; it acquires traces of the adult head and bill pattern by the end of the first winter.

Voice Normally silent, but will utter croaks and other grunting notes. Males in courtship utter a whistling call.

Habitat Tundra and forest bogs in summer; coastal waters during winter.

Range Breeds from Alaska throughout northern Canada to Newfoundland. Winters along both coasts; rare visitor to the Gulf Coast.

144

White-winged Scoter *Melanitta fusca*

The largest of the three scoter species, the White-winged is easily distinguished from other scoters in flight by the large white patches in its secondaries. Like most sea ducks, White-winged Scoters subsist primarily on shellfish, which they pluck off the ocean floor with their stout bills and then swallow whole. Their extraordinarily powerful gizzards are capable of crushing even the hardest shells.

Identification 19–23½". A heavy-bodied bird. Males are black with conspicuous white patches in the secondaries, visible at a considerable distance. At closer range, a small white patch around the eye can be seen. The bill is orange with a black knob at the base. Females are dark brown with white wing-patches as in males. There are two whitish patches on the side of the head. Immature males are like the female.

Voice A hoarse croak.

Habitat Boreal forest and tundra bogs and ponds in summer; seacoasts in winter.

Range Breeds in Alaska and western Canada; winters along both coasts.

Common Goldeneye *Bucephala clangula*

In flight the wings of this handsome duck produce a loud, whistling sound, giving rise to the species' nickname, "whistler." Even at rest, the drake's bold black-and-white plumage renders the bird very conspicuous. This species is one of the few ducks that nest in abandoned woodpecker holes in trees.

Identification 16–20". The bright yellow eye for which this duck is named is visible at close range. Males are mostly bright white with a black back and a dark, iridescent green head. There is a bold, circular white patch on the face, forward of the eye. Females and immature males are mostly gray with a paler belly and dark brown head. Large white wing-patches on the inner half of the wing are visible in flight.

Voice Courting males produce a shrill *jeee-ep;* females, a low quack.

Habitat Summers on lakes and bogs in coniferous forests; winters on coastal harbors and bays as well as inland waters.

Range Breeds from Alaska throughout Canada to Newfoundland and the northern U.S. Winters in the U.S. on open waters.

Barrow's Goldeneye *Bucephala islandica*

Barrow's Goldeneye is much less common than its close relative the Common Goldeneye (see page 148). This species is a cavity-nesting duck, often choosing an abandoned woodpecker hole in which to raise its young. The male Barrow's Goldeneye engages in elaborate courtship displays, including one in which he alternately dips his head forward and arches it backward over his back.

Identification 16½–20". Males are generally dark above and light below. The crescent-shaped white mark in front of the eye contrasts with the dark head. (The Common Goldeneye sports a round white mark.) Females have a dark head, white collar, and grayish body; bill is sometimes yellow.

Voice Mewing and croaking notes during breeding; otherwise silent.

Habitat Forested lakes and rivers; also coastal waters in winter.

Range Breeds from Alaska south to Oregon and Montana, also coastal Labrador; winters along the Pacific Coast from Alaska to California, and from Newfoundland to Long Island, New York.

Bufflehead *Bucephala albeola*

This dapper little bird is the smallest duck in North America. Its proportionally large head gives rise to its common name, which is derived from its scientific name, *Bucephala*, meaning buffalo-headed. Like its larger relatives the goldeneyes, the Bufflehead nests in abandoned woodpecker holes in trees.

Identification 13–15½". A very small duck with a proportionally large head. The male is bold black and white with an iridescent green-and-purple head that has a white patch across the nape. Females and immature males have a gray breast and belly; they are otherwise dark brown with a white spot behind the eye. In flight, birds show a small white wing-patch in the secondaries.

Voice Males give a squeaky whistle; females, a low quack.

Habitat Boreal forests near small lakes and ponds in summer; saltwater bays in winter; also large lakes and rivers.

Range Breeds from central Alaska to Quebec, and in Washington, Idaho, Montana, and Wyoming. Winters along all three coasts and inland, except in the southern states.

Smew *Mergellus albellus*

The Smew is a Eurasian species that occurs with some regularity in the Aleutians but is a rare vagrant elsewhere in North America. It is another of the waterfowl species raised domestically by duck fanciers, so sightings outside its normal range may be of domesticated individuals. The Smew is the smallest of the mergansers, but in contrast to most mergansers it eats relatively few fish. It is also another of the cavity nesters and may be found in summer in forested areas of Russia and Siberia.

Identification 14–16". Males are largely white with various black lines on the body and head. There is an extensive black eye-patch and a petite black bill. The female's head is rusty red on top and white below; her body is mottled grayish brown.

Voice Normally silent; males utter weak whistling notes.

Habitat Freshwater streams and ponds; also coastal waters in winter.

Range A Eurasian species that may be seen in fall and winter off the western and central Aleutian Islands.

Hooded Merganser *Lophodytes cucullatus*

Mergansers are typical diving ducks, characteristically feeding below the surface of the water. While the Hooded Merganser is an excellent underwater swimmer and takes its share of fish, it also feeds on crustaceans and insects. In late winter and early spring, it is often possible to observe their interesting courtship displays. The drake typically shows off his handsome crest and then arches his neck quickly backward so that his head nearly touches his back.

Identification 16–19". The drake is buff below with a white chest and black back and head. The contrasting white head patch and crested appearance are diagnostic. Females are light brown with a pale crest.

Voice Hoarse grunts and croaks.

Habitat Woodland ponds, lakes, and rivers, and sometimes saltwater creeks in winter.

Range Breeds from southeasternmost Alaska south to Oregon, and from Manitoba east to Nova Scotia south to Arkansas. Winters on the California coast, and from the Gulf Coast northward to New England.

156

Common Merganser *Mergus merganser*

True to the Latin name *mergus,* meaning diver, the Common Merganser is an underwater hunter whose diet consists mainly of fish. Once below the surface, this duck uses its expert swimming abilities to chase down minnow-sized fish. The merganser's bill is equipped with a toothlike edge that helps the bird hold onto its prey. This is the largest of the three North American merganser species (the others are the Hooded and the Red-breasted mergansers) and the largest wild duck likely to be seen inland.

Identification 22–27". A large duck. The drake is recognized by his white breast, green head complete with crest feathers, and thin red bill. The female has a reddish-brown head and crest and a contrasting white throat area.

Voice Harsh, rasping croaks.

Habitat Usually inland, freshwater ponds and wooded rivers; in winter, occasionally found on salt bays.

Range Widespread from eastern Alaska throughout Canada and the U.S. Winters throughout much of the U.S.

Red-breasted Merganser *Mergus serrator*

If ducks are best defined by their bills, then mergansers are the least ducklike of all ducks. One of three mergansers in North America, the Red-breasted shares with its relatives a long, forceps-shaped, serrated bill that is superbly adapted for catching fish. The Red-breasted differs in being almost exclusively marine when not breeding.

Identification
19½–26". A duck with a long, narrow, bright red bill. Adult males have a black back, gray sides, reddish-brown speckled breast, and a white collar. The dark iridescent green head carries a wispy crest that extends backward. Females and immature males are largely grayish brown with a crested, rusty-brown head. In all plumages, the white on the inner half of the wing is visible in flight.

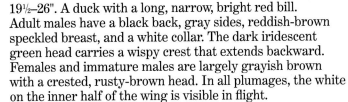

Voice
On breeding grounds, males utter a soft *yeow-yeow;* females give a harsh *karrr.*

Habitat
Lakes, rivers, and occasionally sheltered coasts in summer; usually winters on salt water.

Range
Breeds from Alaska east throughout northern Canada as far south as the Great Lakes; winters along all three coasts.

160

Ruddy Duck *Oxyura jamaicensis*

Something of an oddball among North American waterfowl, the Ruddy Duck belongs to a group called "stiff-tailed" ducks. In breeding plumage, the chunky drake, complete with cocked tail, appears to be something of a caricature. Ruddy Ducks spend most of their time on the water and tend not to associate with other waterfowl. Expert divers, they gather food on the bottoms of ponds and lakes.

Identification | 14–16". A small, chunky duck. Drakes in breeding plumage are rust-colored with a broad, pale blue bill and white cheek. Females are overall light brown, with the distinctive bill and body shape. They also show a single cheek stripe.

Voice | Usually silent except in courtship, when males utter continuous clucking and *chuck* notes.

Habitat | Freshwater marshes, lakes, and ponds with dense vegetation; in winter, on large bodies of water and occasionally on salt marshes.

Range | Breeds from British Columbia to Manitoba, south to Texas, and in the Northeast. Winters around the coastal perimeter of the U.S. and as far inland as Missouri.

162

Masked Duck *Oxyura dominica*

The Masked Duck is a tropical member of the "stiff-tailed" duck group. These ducks, including the more common Ruddy Duck, reportedly use their rigid tail feathers to help steer them when swimming underwater. Masked Ducks employ an unusual submarine approach when ready to fly from a pond. Typically the duck makes a brief dive, surfaces, and launches vertically into the air.

Identification 12–14". A small duck. The male is overall reddish brown with a white speculum. In breeding plumage he has a distinctive black face and a partially blue bill. Females (one is shown here), winter males, and juveniles are largely mottled brown with a characteristic pair of dark facial stripes, one through the eye and one across the cheek.

Voice Males call a distinctive *kuri-kuroo, kuri-kroo;* females cluck and hiss.

Habitat Dense marshes, but may be seen on open water.

Range Rare visitor to the Gulf Coast of Texas and Florida. Several records along the Atlantic coast to Pennsylvania.

Purple Gallinule *Porphyrula martinica*

Gallinules, chickenlike members of the rail family (Rallidae), are closely related to cranes. The adult Purple Gallinule's spectacular plumage makes it one of the more attractive water birds. While its normal range in the United States is restricted to the Southeast, the Purple Gallinule has a reputation for showing up in unexpected places—sometimes as far north as southern Canada. Such seasonal wandering or vagrancy is known to occur in many of North America's birds, and it adds spice to the pastime of birdwatching. While surprises are possible at any time, most vagrancy tends to be noted in the fall.

Identification   11–13". A chicken-sized marsh bird with characteristically large feet. It is blue below and green above; its red bill has a yellow tip. Juvenile birds show some green and blue coloring but are largely pale brown.

Voice   Henlike cackling, squawks, and grunts.

Habitat   Freshwater swamps and marshes.

Range   Breeds in the southeastern U.S. and the Gulf Coast inland to Texas. Winters in Florida, and along the Gulf Coast.

166

Common Moorhen *Gallinula chloropus*

Formerly called the Common Gallinule, this species is actually uncommon but fairly widespread in eastern North America. The legs and feet of gallinules are especially well adapted for life in the marsh. Common Moorhens can actually climb through dense, reedy vegetation by grasping first one stalk and then another. Their long toes distribute their weight, allowing them to walk over muddy ground and across lily pads and other broad-leaved water plants. The Common Moorhen's diet is largely vegetable; on occasion, however, these birds take small crustaceans and insects.

Identification 13". A chicken-sized marsh bird that is mainly gray-black with a red shield on its forehead and a yellow-tipped red bill. It has a white rump and a narrow white line on the flanks.

Voice A wide variety of harsh squawks, henlike clucks, and screams.

Habitat Freshwater marshes, ponds, and edges of rivers with cattails, willows, and other vegetation.

Range From California south to New Mexico, and from Oklahoma east to southern New England and south to the Gulf Coast.

American Coot *Fulica americana*

Technically a member of the rail family, the American Coot behaves more like a waterfowl. While its close relatives the gallinules and rails tend to stay in heavy cover or along edges, the coot regularly swims in open water. There it combines the strategies of both diving and dabbling ducks to feed on a variety of aquatic plants and arthropods. The American Coot is easily distinguished from the ducks, however, when it decides to become airborne. Many ducks seem to spring from the surface, but the coot patters along with flailing wings in an apparently labored takeoff.

Identification   15". An overall chunky, slate-gray, chickenlike bird with a white bill and white undertail feathers. It often bobs its head when swimming. In flight, a white trailing edge on the wing is seen.

Voice   A variety of clucks, cackles, and grunts.

Habitat   Freshwater ponds, lakes, and open marshes; winters on both fresh and salt water.

Range   Widespread in southern Canada and throughout the U.S. Winters mainly in the southern U.S.

170

WATERFOWL
IN FLIGHT

Loons in Flight

While all loons spend much of their time on the water, they are also likely to be observed in flight. At any appreciable distance, loons (and for that matter, all birds) will appear more or less as dark silhouettes. When in flight the loon carries its head and neck and legs and feet outstretched. These extremities often droop a bit, giving the bird a hunched appearance. Loons keep up a constant stroking motion with their wings and seldom glide. The relative size of the head and body as well as the speed of the wing beat may help distinguish the species. Migrating loons are more likely to be seen individually, or in pairs or small groups, than in large flocks.

Pacific Loon

Cormorants in Flight

The four species of cormorants covered in this guide are often observed in flight. Great, Pelagic, and Neotropic cormorants fly with head and neck outstretched. (Double-crested Cormorants draw the head and neck in slightly.) In none of these species do the feet extend past the tail. Cormorants may sometimes be observed in soaring flight, spiraling upward in large circles. Migrating cormorants often travel in large, V-shaped wedges or long, irregular lines, similar to those of migrating geese, but they are always silent. From below, immature Double-crested Cormorants and Great Cormorants can often be separated by the light and dark areas of their underparts.

Double-crested Cormorant

Swans in Flight

The four species of swans included in this guide are several of North America's largest birds. Because of its massive size, the swan's flight often seems ponderous and labored. The wing beats are typically deep and relatively slow. The long, outstretched neck and head create a good field mark. Large herons and egrets normally fly with their necks tucked back. Cranes fly with long, outstretched neck and head, but their legs and feet extend well past the tail. In contrast, the swan's legs and feet do not extend past the tail. Migrating swans regularly travel in loose lines or V's.

Trumpeter Swans

Geese in Flight

Geese congregate in large flocks as they fly to and from feeding and resting areas and while on migration. In profile, most geese present a heavy-bodied appearance. The neck and head are outstretched, but the feet do not extend past the tail. Along with their typical, V-shaped flight formations, their vocalizations are useful clues in determining the nature of the flock and even the species involved. Geese keep up a constant flapping motion when airborne, except when preparing to land.

Snow Geese

Ducks in Flight

In general, the duck species covered in these pages are smaller than the other birds shown. Their wing beats, therefore, will appear relatively more rapid. Many of the dabbling ducks, such as American Black Ducks, Mallards, and the various teal species, typically travel in fairly small groups; while some of the diving ducks, such as eiders, scoters, and mergansers, may fly in medium- to large-sized flocks. Among the variety of species covered here, flight characteristics will vary. For example, teal may be recognized on the wing by their small size and rapid flight. Common Goldeneyes can be identified by a distinct whistle created by the movement of air across their flight feathers. The mergansers fly in a typically "flat" posture, with the body and wings apparently held in the same plane. Each duck species also has a characteristic light and dark pattern created by its plumage. Individual species descriptions give these details.

Northern Pintails

Parts of a Bird

crown

forehead

throat

breast

speculum

flank

belly

nape

back

scapulars

tertials

rump

primaries

tail feathers

undertail coverts

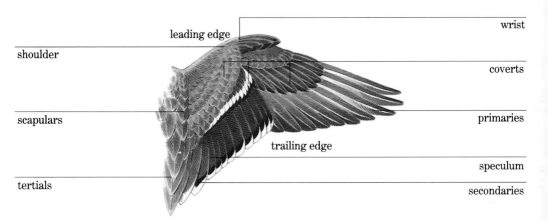

wrist

leading edge

shoulder

coverts

scapulars

primaries

trailing edge

speculum

tertials

secondaries

185

Glossary

Accidental
(See Vagrant.)

Arthropod
An invertebrate animal with jointed body and limbs, such as an insect, arachnid, or crustacean.

Casual
Occurring outside its normal range but somewhat more frequently than a vagrant.

Clutch
A set of eggs laid by one bird.

Coverts
The small feathers covering the bases of other, usually larger, feathers. Coverts provide a smooth, aerodynamic surface.

Crest
A tuft of elongated feathers on the crown.

Crown
The uppermost surface of the head.

Eye-ring
A fleshy or feathered ring around the eye.

Eye-stripe
A stripe running horizontally from the base of the bill through the eye.

Flight feathers
The long feathers of the wing and tail used for flight. The flight feathers of the wing are composed of primaries, secondaries, and tertials.

Lore
The area between the base of the bill and the eye.

Mandible
One of the two parts (upper and lower) of a bird's bill.

Mantle
The back and the upper surfaces of the wings.

Mask
An area of contrasting color on the front of the face and around the eyes.

Morph
One of two or more distinct color types within the same species, occurring independently of age, sex, season, and geography. Also referred to as a phase.

Nape
The back of the head and the hindneck.

Phase
(See Morph.)

Race
(See Subspecies.)

Rump
The lower back, just above the tail.

Scapulars
A group of feathers on the shoulder, along the side of the back.

Speculum
A distinctively colored area on the trailing edge of the wing, especially the iridescent patch on the secondaries of some ducks.

Subspecies
A geographical population that is slightly different from other populations of the same species. Also called a race.

Taiga
The belt of coniferous forest covering the northern part of North America and Eurasia from coast to coast.

Underparts
The lower surface of the body, including the chin, throat, breast, belly, sides, and undertail coverts.

Vagrant
A bird that occurs outside its normal range. Also referred to as accidental.

Wing bar
A bar of contrasting color on the upper wing coverts.

Wing lining
A collective term for the coverts of the underwing.

Wing stripe
A lengthwise stripe on the upper surface of the extended wing.

Index

189

Credits

Photographers

Ron Austing (153)
Eliot Cohen (167)
Rob Curtis/The Early Birder (51)
Richard Day (79)

DEMBINSKY PHOTO ASSOCIATES:
Carl R. Sams (29)
Roger Wilmshurst (113)

Larry R. Ditto (53)
Jeff Foott (105, 131, 157, 179)
Chuck Gordon (33, 37, 39, 67, 107, 111, 121, 147, 149)
John Heidecker/Nature Photos (75)
Robert Y. Kaufman/Yogi, Inc. (47, 151)
Kevin T. Karlson (27)
G.C. Kelley (59, 69, 85, 163)
Harold Lindstrom (119, 129, 143)
Bates Littlehales (35, 49, 55, 123, 145)
Charles W. Melton (109)
Arthur & Elaine Morris/
Birds As Art (3, 22-23, 31, 41, 43, 45, 95, 101, 115, 171, 172-173, 181)

PHOTO/NATS, INC.:
Cortez C. Austin, Jr. (71)
Priscilla Connell (93)

Rod Planck (159)
Robert & Jean Pollock (103, 155)
Betty Randall (25, 133)
James H. Robinson (177)

ROOT RESOURCES:
Kenneth W. Fink (81)
Jim Flynn (137, 141)

Ron Sanford (183)
Gregory K. Scott (83)
Robert C. Simpson (169)
Frank S. Todd (57, 63, 65, 89, 117, 135, 175)
Tom J. Ulrich (73, 77, 87, 125, 139)

VIREO:
W. S. Clark (99)
B. Gadsby (97)
P. W. Sykes, Jr. (165)

Mark F. Wallner (161)
Larry West (91)
Tim Zurowski (Front Cover, 61, 127)

Cover Photograph: Mallard by Tim Zurowski
Title Page: Clarke's Grebe by Arthur & Elaine Morris/Birds As Art
Spread (22-23): Snow Geese by Arthur & Elaine Morris/Birds As Art
Spread (172-173): Snow Geese by Arthur & Elaine Morris/Birds As Art

Illustrators

Range maps by Paul Singer
Drawings by Barry Van Dusen (184-185)
Silhouette drawings by Douglas Pratt and Paul Singer

The photographers and illustrators hold copyrights to their works.

Staff

This book was created by
Chanticleer Press.
All editorial inquiries should
be addressed to:
Chanticleer Press
568 Broadway, Suite #1005A
New York, NY 10012
(212) 941-1522

Chanticleer Press Staff
Founding Publisher:
Paul Steiner
Publisher: Andrew Stewart
Managing Editor: Edie Locke
Production Manager:
Deirdre Duggan Ventry
Assistant to the Publisher:
Kelly Beekman
Text Editor: Carol M. Healy
Consultant: John Farrand, Jr.
Photo Editor: Lori J. Hogan
Designer: Sheila Ross
Research Assistant:
Debora Diggins

Original series design by
Massimo Vignelli.

To purchase this book, or other
National Audubon Society
illustrated nature books,
please contact:
Alfred A. Knopf, Inc.
201 East 50th Street
New York, NY 10022
(800) 733-3000